Great White Shark Migration

By Susan H. Gray

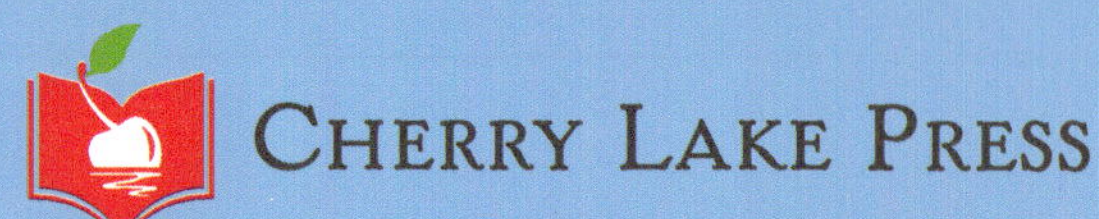

Published in the United States of America by Cherry Lake Publishing Group
Ann Arbor, Michigan
www.cherrylakepublishing.com

Reading Adviser: Marla Conn, MS, Ed., Literacy specialist, Read-Ability, Inc.
Content Adviser: Sheila K. Schueller, Ph.D., Lecturer and Academic Program Specialist, University of Michigan School for Environment & Sustainability
Photo Credits: ©wildestanimal/Shutterstock.com, cover; ©Joe Belanger/Shutterstock.com, 4; ©Willya Bradberry/Shutterstock.com, 6; ©Fer Gregory/Shutterstock.com, 8; ©Fiona Averst/Shutterstock.com, 10; ©Krzysztof Odziomek/Shutterstock.com, 12; ©atese/iStock.com, 14; ©Sergey Uryadnikov/Shutterstock.com, 16; ©Andrea Izzotti, 18; ©Marc Henauer/Shutterstock.com, 20

Cherry Lake Press is an imprint of Cherry Lake Publishing Group.

Library of Congress Cataloging-in-Publication Data

Names: Gray, Susan Heinrichs, author.
Title: Great white shark migration / by Susan H. Gray.
Description: Ann Arbor, Michigan: Cherry Lake Publishing, [2021] | Series: Marvelous migrations | Includes index. | Audience: Grades 2-3
Identifiers: LCCN 2020002791 (print) | LCCN 2020002792 (ebook) | ISBN 9781534168602 (hardcover) | ISBN 9781534170285 (paperback) | ISBN 9781534172128 (pdf) | ISBN 9781534173965 (ebook)
Subjects: LCSH: White shark–Migration–Juvenile literature.
Classification: LCC QL638.95.L3 G718 2021 (print) | LCC QL638.95.L3 (ebook) | DDC 597.3/31568–dc23
LC record available at https://lccn.loc.gov/2020002791
LC ebook record available at https://lccn.loc.gov/2020002792

Cherry Lake Publishing Group would like to acknowledge the work of the Partnership for 21st Century Learning, a Network of Battelle for Kids. Please visit http://www.battelleforkids.org/networks/p21 for more information.

Printed in the United States of America
Corporate Graphics

CONTENTS

Great whites got their name from their white underbellies.

The Middle of Nowhere

It's January and great white sharks are absolutely stuffed. Seals had been plentiful along California's coast. And the great white **predators** enjoyed the feast.

Soon, the sharks will head out into the Pacific Ocean. They will **migrate** west to a spot that scientists discovered only recently. There, in the middle of nowhere, sharks will hang around for months.

Great whites can smell blood from up to 3 miles (5 kilometers) away.

Experts never thought of the open ocean as packed with **prey** animals. It seemed that most sea life was closer to land. But around 20 years ago, California scientists tracked some great whites. The sharks left the coast and swam halfway to Hawaii. Then, they just stopped. It turned out that they were feeding in this open area. Scientists were stumped. Somehow, there was enough prey to feed a gang of 4,000-pound (1,814-kilogram) predators! The scientists called the region the "**White Shark Café**." And they decided to learn more.

In movies, a shark's exposed dorsal fin is always a sign of danger. In real life, sharks rarely swim on the surface.

Minds of Their Own

It can be hard for scientists to track the migration of great whites. The sharks often have minds of their own.

Make a Guess!

Imagine you are a scientist who studies sharks. What are some problems you might have with following their migrations?

Female great whites are larger than males.

Sharks near Australia may travel similar paths to their **destinations**. But they arrive and depart at different times of the year. Sharks living at the southern tip of Africa migrate north along the coast. Others swim east for more than 5,400 miles (8,690 km) to Australia.

Think!

Some experts say these sharks can live to be 30 years old. Some say 40 or 50 years old. Why is it so hard to figure this out?

The whale shark is much bigger than a great white. It can grow up to 60 feet (18 meters) long.

The White Shark Café

So what about those California sharks out in the Pacific? It turned out they really did find a café. Fish, jellyfish, and squid were out there in **droves**. This really surprised the scientists. To find out more, they knew they had to look deeper. Deeper in the ocean, that is.

Scientists found that these prey animals have daily migrations of their own. Before

Divers can observe sharks safely while cage diving.

dawn, they begin moving down to deep, dark, safe waters. By nightfall, they are moving back up again.

Scientists placed special tags on a few sharks at the café. The tags kept track of how deep the sharks were diving. The scientists learned that the great whites were following their prey. In the daytime, sharks were diving almost 1,500 feet

Look!

Look at some pictures of the great white shark's body. What are some features that make this shark a great swimmer?

Great whites will surprise their prey by jumping out of the water and catching their meal in their mouth.

(457 m) deep. At night, they went down only 650 feet (198 m).

For the prey, the deep, dark water wasn't so safe after all. Maybe the sharks could not see them, but that didn't matter. Great whites can locate their prey in other ways. They can sense faint electrical charges. So a fish's beating heart might alert a shark. The smell of one drop of blood in the water can **betray** a wounded seal. And great whites can sense the movement of animals in the water up to 820 feet (250 m) away!

Great white sharks do not migrate in groups.

More and More Questions

Sharks stay at the café for weeks or even months. Then, some swim on to Hawaii. Others move to a second café south of Hawaii. Still, others migrate right back to California. Seals will be arriving along the state's coast to spend the fall and winter there. Great whites will not be far behind.

The more scientists learn about great white shark migrations, the more questions they

Humans have only 32 teeth. Sharks can have as many as 300!

have. Why do some sharks skip the café altogether? How come males migrate every year, while females migrate every 2 years? And could there be more white shark cafés out there? Scientists will be studying these fish and their migrations for many years. We still have much to learn.

Ask Questions!

The **liver** of an adult white shark is huge. At the beginning of a migration, the liver is packed with stored fat. Fat provides energy for the trip. Where does the shark get all that fat?

GLOSSARY

betray (bih-TRAY) to reveal or expose

destinations (des-tuh-NAY-shuhnz) the places that animals or people are traveling to

droves (DROVEZ) large numbers

liver (LIV-ur) the organ in an animal that produces a fluid that helps digest food

migrate (MYE-grayt) to move from one region to another and back again

predators (PRED-uh-turz) animals that hunt and eat other animals

prey (PRAY) animals that are eaten by other animals

White Shark Café (WHITE SHARK kah-FAY) a feeding ground in the middle of the Pacific Ocean

FIND OUT MORE

BOOKS

Roy, Katherine. *Neighborhood Sharks: Hunting with the Great Whites of California's Farallon Islands*. New York, NY: David Macaulay Studio, Roaring Brook Press, 2014.

Schreiber, Anne. *Sharks!* Washington, DC: National Geographic Children's Books, 2008.

Worth, Bonnie. *Hark! A Shark! All About Sharks*. New York, NY: Random House Books for Young Readers, 2013.

WEBSITES

ActiveWild—Great White Shark Facts

https://www.activewild.com/great-white-shark-facts-for-kids
Watch a video showing how sharks are tagged and also read fun facts about great whites.

National Geographic Kids—Great White Shark

https://kids.nationalgeographic.com/animals/fish/great-white-shark
Learn more about great whites, their babies, and how they hunt.

INDEX

ABOUT THE AUTHOR

Susan H. Gray has a master's degree in zoology. She has written more than 170 reference books for children and especially loves writing about animals. Susan lives in Cabot, Arkansas, with her husband, Michael, and many pets.